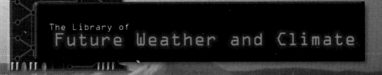

The Library of
Future Weather and Climate

Ice Ages
of the Future

Paul Stein

The Rosen Publishing Group, Inc.
New York

Published in 2001 by The Rosen Publishing Group, Inc.
29 East 21st Street, New York, NY 10010

Copyright © 2001 by The Rosen Publishing Group, Inc.

First Edition

Library of Congress Cataloging-in-Publication Data

Stein, Paul, 1968–
Ice ages of the future / Paul Stein. — 1st ed.
p. cm. — (The library of future weather and climate)
Includes index.
ISBN 0-8239-3415-2
1. Glacial epoch—Juvenile literature. 2. Climatic changes—Juvenile literature. 3. Nature—Effect of human beings on—Juvenile literature.[1. Glacial epoch. 2. Climatic changes. 3. Nature—Effect of human beings on.] I. Title. II. Series.
QE697 .S785 2001
551.6—dc21

00-011433

All temperatures in this book are in degrees Fahrenheit, except where specifically noted. To convert to degrees Celsius, or centigrade, use the following formula:

Celsius temperature = (5 ÷ 9) x (the temperature in Fahrenheit - 32)

Manufactured in the United States of America

Contents

Introduction

The vast ice sheet that covers Greenland is one of the most desolate regions of the world. Extending over 1.1 million square miles of land on the world's largest island, the Greenland ice sheet contains nearly 10 percent of the earth's freshwater supply locked up in frozen form. On top of the ice sheet, summertime temperatures rarely exceed freezing, while in the winter, temperatures can plunge to sixty below zero. Occasional blizzards whip snow across the icy landscape. Here, far above the Arctic Circle and hundreds of miles from the nearest settlement, a group of American scientists worked each summer from 1989 to 1993 looking for clues about the earth's climate.

Drilling deeper and deeper into the ice, as many as sixty researchers worked in the unforgiving, alien environment. Supplies had to be flown in on planes equipped with skis for landing on the ice. Snow was melted to collect water. At an elevation of over 10,000 feet, the thin air made physical work extremely tiring. Occasional whiteouts, periods of driving, stinging, wind-blown snow, obscured visibility. Strange, shimmering mirages, called fata morgana, sometimes gave rise to icy phantom mountains along the horizon. And for months the scientists had no nighttime darkness to help them sleep. This far north the Sun does not set from early May through early August.

The scientists drilled down into the ice to learn about the air above them and how it has changed over thousands of years. Undisturbed ice, such as the ice that covers most of Greenland, provides a remarkable record of the atmosphere through history. Each year, around two to three feet of snow falls on central Greenland. Year after year, new layers of snow are deposited on the surface, pressing down on the older snow below and compacting it into ice. This process has been going on for so long that the Greenland ice cap is five miles thick in places. Tiny air bubbles trapped in the ice provide a record of the earth's air at the time the ice formed. By drilling into the ice, scientists are able to extract ice cores—long narrow tubes of ice that allow scientists to look back in time. Examining the air bubbles in the layers set down each year, scientists can learn about the earth's climate through history. The deeper the scientists drill, the further back in time they can peer.

In 1993, the drilling ended as scientists struck the Greenland bedrock, the actual island on which the massive ice sheet rests. At over 10,000 feet deep, the ice core was the longest ever retrieved in the

Northern Hemisphere, providing a record of climate 100,000 years old. Using this ice core and a similar one drilled seventeen miles away by a team of European researchers, scientists have been able to unlock some of the secrets of the earth's climate.

Part of this historical record tells of the comings and goings of great ice ages, periods of time thousands of years long when the planet was plunged into much colder conditions, and vast ice sheets spread toward the equator. Examining ice core data, scientists can clearly see the changes in the atmosphere that occurred during the peak of the last ice age some 20,000 years ago. More important, they have seen startlingly clear evidence that the earth's climate has not always changed gradually over time as had been previously thought. The Greenland ice record tells of global cold snaps, drastic and disruptive changes in the earth's climate, occurring thousands of years ago over a period of mere decades.

Looking back through time using ice core data, ice ages and cold snaps seem to be a normal part of our planet's changing and volatile climate. This book investigates the processes that lead to an ice age. Ice ages may seem a remote concern to us today, especially as global warming becomes more and more of a reality. But the same natural processes that play a role in global warming are also linked to the ice ages and the rapid cold snaps that scientists have learned about by studying ice core data. Ice ages of the future, lasting tens of thousands of years or longer, are a distant reality. Unexpected global cold snaps, occurring well within a person's lifetime, could happen much sooner.

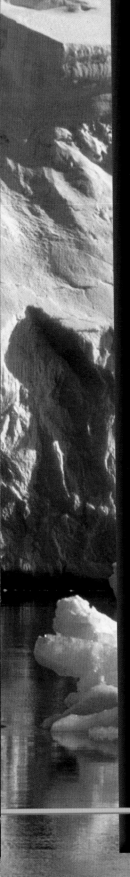

1 Ice Ages of the Past

Twenty thousand years ago, the continent of North America was vastly different than it is today. Dinosaurs had been extinct for millions of years. Sea levels were as much as 400 feet lower, exposing over 14 million square miles of land worldwide. Some of this exposed land included a natural causeway between North America and Asia, stretching between modern-day Alaska and Siberia. It would be thousands of years until the first humans, ancestors of the Native Americans, would cross the land bridge from Asia before rising oceans covered it up again.

Much of the world's water is locked up in the frozen ice caps in the northernmost and southernmost regions of the globe. These ice caps have grown and shrunk over millions of years.

The reason the oceans were so low at that time was because much of the earth's water was locked up in enormous glaciers. A vast sheet of ice, two miles deep in some places, covered most of what is now Canada, extending southward into the northern United States. Where Chicago and New York City stand today, a mountainous glacier smothered the land 20,000 years ago. Another giant ice sheet expanded from Scandinavia into parts of Asia and Europe, covering most of what is now the British Isles. The northern Atlantic was choked with ice, with icebergs probably floating as far south as modern-day Portugal.

A large part of the planet was plunged into frigid conditions. Globally, the average temperature was some five to nine degrees lower than today. While this may seem like a minor cool-down, in planetary terms this kind of temperature drop is catastrophic. Over parts of North America the climate was even colder, similar to the bitter conditions experienced in Greenland today.

Since these global cold waves came and went long before humans gained the ability to record history, how

Geological patterns in mountains and rock leave clues about the historical progress of ice ages.

do we know they actually occurred? One way is by examining the land. People in the Alps have long observed that the advancing and retreating mountain glaciers leave telltale scars on the earth, transporting rocks and debris down from the mountainsides and depositing them in mounds along the leading edge of the ice. Glaciers can also wrench giant boulders off a mountain and slowly carry them downslope over decades or centuries, depositing them on land far from their original source. And as a massive glacier crawls over rocks, it can scour them, much like the action of sandpaper on wood, leaving behind characteristic scratching marks. As geologists studied the land across northern Europe and North America, they began to see many of these same kinds of features, far from any mountain glaciers. The only explanation was the presence of vast ice sheets thousands of years ago.

Evidence for ice ages also comes from the study of past climates. This area of scientific research is known as paleoclimatology, a word that comes from the Greek word *paleo*, meaning "ancient." Paleoclimatologists use a variety of different methods to understand how the climate of the earth has changed over millions of years. These methods include the use of ice cores, the study of tree rings, the drilling of cores in the bottoms of lakes and oceans, and the examination of coral.

We've already seen how scientists in Greenland drilled a 100,000-year-old record of climate from the ice. Another ice core has been extracted from Vostok, Antarctica, yielding a 200,000-year-old record of the atmosphere. Scientists study the oxygen molecules trapped in the ice at the time

Studying the thickness of tree rings helps scientists figure out how the climate has changed over many hundreds of years.

it formed. Atoms of oxygen come in two main varieties, called isotopes, that differ slightly in their weight. Water molecules, consisting of two atoms of hydrogen and one of oxygen, can contain either kind of oxygen isotope. When ocean water evaporates into the air, the heavier oxygen isotope gets left behind in the water more often. The relative number of heavier and lighter oxygen isotopes that evaporate depends on the air temperature. By studying the relative amounts of the different isotopes in ice cores, scientists are able to estimate the temperature of the air at the time the ice was formed. Knowing the air temperature gives scientists valuable information about the climate as a whole.

Scientists also drill cores in the bottoms of lakes and oceans. Instead of studying ice, however, the scientists examine the layers of debris that fall to the bottom of the lake or ocean over the years. This debris includes microscopic shells and plant life, dust and pollen from the air, and the skeletons of fish. In the same way that falling snow forms layers of ice on Greenland, the slow rain of debris onto the bottoms of lakes and oceans causes thin layers of sediment to build up over time. Scientists study the chemical com-position of these lake and ocean bottom cores to determine what the climate of the earth was like when the sediment formed. For example, scientists can pick out tiny, fossilized pollen grains from the sediment. Different kinds of pollen are produced from different plants, and each kind of pollen has a distinct shape. So when sci-entists find pollen in ocean or lake bottom cores, they can tell what kinds of plants were living, and in what abundance, at the time the pollen was deposited. The kinds of plants that live during a given time depend on the climate, so studying pollen records can yield valuable information on the earth's atmosphere long ago.

Coral can also provide important information on the earth's climate. Coral grows using a substance called calcium carbonate, which it gets from ocean water. Calcium carbonate contains oxygen isotopes. Just as with ice cores, scientists study the relative amounts of the different isotopes in coral to learn about the earth's temperature at the time the coral grew. Fossilized coral, living many thousands of years ago, sometimes becomes exposed if the local terrain rises slowly through the years. Such a record of ancient climate comes from exposed coral reefs in New Guinea and Indonesia.

Paleoclimatologists also study tree rings to learn about regional climate conditions over hundreds or thousands of years. The differing thicknesses of tree rings are influenced by the varying temperature and rainfall for each year of a tree's life. The older the tree, the further back in time a paleoclimatologist can probe to learn about regional weather patterns.

Scientists, therefore, have several different ways of learning about the climate of the earth long before humans began keeping records. The evidence tells of numerous periods of glaciation that chilled the planet over thousands of years. The ice age that gripped the planet 20,000 years ago was by no means unprecedented.

Other ice ages have come and gone throughout prehistory. Scientists think that one of the coldest episodes in the earth's history occurred 2.2 billion years ago, when ice sheets may have expanded to within 800 miles of the equator. Severe global cold spells also likely occurred between 500 and 750 million years ago, and between 250 and 350 million years ago. During the last two million years, scientists have counted over twenty ice ages. What causes the planet to

The calcium carbonate in coral gives scientists clues about the earth's climate thousands of years ago.

become so cold that glaciers are able to expand across continents? In the next chapter, we will examine the processes that can lead to an ice age.

2 Setting the Stage for Ice

Outside of the tropics, people are familiar with the seasons as a slow cycle of changing weather, accompanied by the greening of plants in the spring and the dying of plants in autumn. Seasons occur because the earth is tilted on its axis. As the earth races through space at over 66,000 miles per hour, the Northern and Southern Hemispheres alternately face toward and away from the Sun over a period of 365 days. When the Northern Hemisphere points toward the Sun and basks in the heat of summer, the Southern Hemisphere points away from the Sun and shivers in the chill of winter.

The daily warming and cooling of the earth causes temperature cycles in the atmosphere. Temperature changes occur over millions of years, as well.

If we think about it, though, there are other kinds of "seasons." For example, each morning the temperature begins to rise as the earth spins to face the Sun. The Sun's rays warm the ground, which warms the air in contact with it. When the Sun sets in the evening, the ground loses its heat source and cools. From colder at night to warmer during the day and back again every twenty-four hours, it's a seasonal cycle in miniature. Like our yearly seasons, the daily temperature cycle is the result of the earth's relationship with the Sun.

There's yet another variety of "season" on our planet, another kind of progression from warmer to colder and back again. These seasons occur more irregularly and over much longer periods of time. The ice age, a global wintertime that can last for thousands or millions of years, is part of this grand seasonal change. The periods in between ice ages, called interglacials, are like a global summertime that can last equally as long. Like our miniature daily seasons, and our familiar yearly seasons, these grand climatic seasons are linked to the Sun and its relationship to the earth.

The earth's climate is shaped by the amount of energy it receives from the Sun and the way this energy is distributed on the earth's surface. The Sun, a giant sphere of blazing gas over 100 times the diameter of the earth and over 300,000 times as massive, fuels the earth's weather machine. It generates an immense amount of energy through the process of nuclear fusion, whereby the nuclei of hydrogen atoms collide and fuse into helium atoms. The energy released by this process is so tremendous that it generates temperatures of 15 million degrees Fahrenheit at the center of the Sun, and over 10,000 degrees Fahrenheit at the Sun's surface. Some of this energy is emitted into space in the form of invisible electromagnetic waves, called radiation. Traveling at the speed of light, some of this radiation eventually strikes the earth, 93 million miles away.

When solar radiation reaches the earth, several different things can happen to it. Some radiation gets deflected, or scattered, in different directions by tiny particles in the atmosphere, such as air molecules and cloud droplets. Some radiation is reflected back into space. Certain kinds of surfaces and objects, such as snow and clouds, are more likely to reflect radiation than other surfaces. These surfaces are said to have a high albedo, meaning that they reflect a large percentage of the radiation that strikes them. In general, darker surfaces, such as asphalt or forests, have a low albedo. As we will see, albedo plays an important role in the development of ice ages.

Most of the radiation not reflected or scattered away from the earth gets absorbed by the ground, the oceans, or by other objects on the surface of the earth. When the earth absorbs radiation, it gains energy, some of which can go into raising the earth's temperature. Like all objects, however, the earth continuously emits, or gives off, radiation in addition to continuously absorbing radiation. Therefore, the temperature of the earth depends on how much radiation it absorbs compared to how much it emits. Where the Sun strikes the earth more directly, as in the tropics, the earth absorbs more radiation than it emits and becomes warm. Where the Sun strikes the

The earth's regions are heated to different levels because of the varying amounts of the Sun's radiation they absorb.

earth more indirectly, as in the polar regions, the earth absorbs less radiation than it emits and cools down.

The earth is therefore unevenly heated between the tropics and poles because of the imbalance in the amount of absorbed solar radiation. This uneven pattern of heating is further complicated because the oceans, which cover nearly 70 percent of the earth, warm and cool much more slowly than the land. It's this uneven distribution of temperature across the planet that sets the atmosphere in motion. Differences in temperature lead to differences in air pressure, which cause the wind to blow and circulate air around the planet. To make matters more complex, the rotation of the earth adds spin to the moving air. Lakes and oceans add moisture. The end result is a complex, ever changing weather machine, with great masses of clashing cold and warm air, dry and moist air, and great spinning areas of high and low pressure that form and decay as they move for thousands of miles across the face of the earth.

Ice ages are a global change in climate. The climate of the earth is just the average of weather conditions across the earth over a long period of time—usually decades or centuries. Over the course of our lives, the weather can vary drastically from day to day. But the climate, the long-term average of all weather events, changes almost imperceptibly. Yet, as evidence from ice and ocean cores, tree rings, and coral shows, the earth's climate does change significantly over time. The climate has its own seasons, with the ice ages being a sort of climatic winter. What could cause the onset of ice ages?

The idea that has gained the most acceptance among scientists was first proposed by a British scientist, James Croll, in the 1860s. He speculated that long-term changes in the amount of solar radiation reaching the earth might account for the changes in climate that can lead to an ice age. In the 1920s, a Serbian mathematician named Milutin Milankovitch developed these ideas into the theory that bears his name. The Milankovitch theory describes how changes in the earth's orbit around the Sun over tens of thousands of years affect the way solar radiation is distributed between the poles and the tropics, thereby affecting climate and setting the stage for an ice age.

There are three different aspects of the earth's orbit that can change over time. The first is the actual path the earth takes around the Sun. The earth doesn't orbit the Sun in a perfect circle, but rather in a shape called an ellipse. An ellipse looks like a slightly squashed circle, though the elliptical shape of the earth's orbit is so slight that it appears nearly circular. This shape, described by scientists using the term "eccentricity," varies slightly over a period of nearly 100,000 years. Changes in eccentricity affect the distribution of solar radiation between the Northern and Southern Hemispheres. However, the role this plays in climate change is unclear. While variations in the amount of solar radiation reaching each hemisphere are very small, the eccentricity cycle seems to be closely tied to the advance and retreat of ice ages.

The second way the earth's orbit can change over time is through changes in the angle of the earth's tilt. Currently, the earth is tilted on its axis at an angle of around 23.5 degrees, though it can vary over a period of 41,000 years from 21.5 to 24.4 degrees. The tilt of

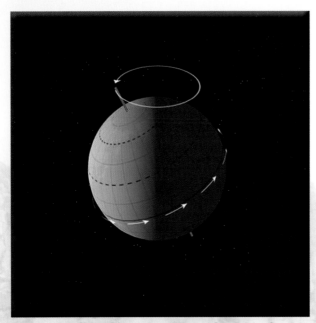

Climate is affected by precession, a change in the orientation of the earth's rotational axis.

the earth reached its maximum value around 9,500 years ago and has been decreasing toward its minimum value ever since. This change in tilt affects the intensity of the seasons. The greater the tilt, the colder the winters and the warmer the summers, on average. The smaller the tilt, the milder the winters and the cooler the summers. Scientists speculate that smaller tilt angles favor the growth of glaciers and ice sheets, since milder winters allow more moisture to gather in the air over the polar regions, leading to heavier snowfall. Cooler summers melt less of the snow that has fallen in the winter.

The third way that the orbit of the earth varies is through a change in the direction of the earth's tilt over a period of about 19,000 to 23,000 years. The effect is similar to the motion of a spinning top that wobbles on its axis. This wobbling motion affects the season of the year during which the earth makes its closest and farthest approaches to the Sun. Currently, the earth is farthest away from the Sun in July and closest in January, but these dates are shifting forward by 1.4 days every 100 years. When the earth is closest to the Sun during summer and farthest away during January, more wintertime snow is able to build up in the Northern Hemisphere.

It's not the individual cycles that are important in the making of an ice age but the combined effects of all three. Since each cycle lasts for different periods of time, the sum of their effects varies. Sometimes the cycles coincide, leading to great changes in the distribution of solar radiation over the planet. Sometimes the cycles tend to cancel one another out. Scientists estimate that orbital changes can reduce the amount of sunshine at high latitudes, closer to the poles, by 10 to 15 percent. The amount of solar radiation received at high latitudes is critical to the development of ice ages because this is where glaciers begin to grow. The colder the polar regions become, the more snow and ice can accumulate.

However, the Milankovitch theory does not completely account for the ice ages. Scientists estimate that changes in the earth's orbit can alter the average temperature of the earth by a few degrees Fahrenheit. That's significant, but it's not enough to cause vast ice sheets to expand across the continents. Other natural processes are needed to deepen the chill.

3 Ice, Oceans, and Land

The Gulf Stream is a great current of warm ocean water that flows from the Gulf of Mexico, around the southern tip of Florida, then northeastward for thousands of miles into the North Atlantic Ocean. Hundreds of miles wide and drifting along at anywhere from one to four miles per hour, the Gulf Stream carries as much as a billion cubic feet of water per second. That's well over a thousand times the rate of flow in the Mississippi River.

But the Gulf Stream does more than move water around. It's part of a system of currents that flows around the entire earth, both on the surface and at the bottom of the oceans. These currents carry warm water out of the tropics and toward the poles, and carry cold water from the poles back toward the tropics. In this way, they act to distribute heat around the earth, and in so doing play an important role as shapers of the earth's climate and contributors to the earth's ice ages.

Ocean water helps determine climate by changing the temperature of the air above it. Cold ocean water, for example, causes the atmosphere above to cool. Likewise, warm ocean water tends to heat the air above. The Gulf Stream, a warm ocean current, can be as much as twenty degrees warmer than the surrounding Atlantic waters. This warm water, originating in the Gulf of Mexico, ends up off the coast of northern and western Europe. After traveling such a long distance, the temperature of the water in the Gulf Stream is not as high as it was in the tropics where it started. However, it's still warm enough to raise air temperatures across much of Europe. Compare, for example, the climates of Paris, France, and Montreal, Canada. Montreal experiences frigid winter temperatures, often not much higher than ten degrees above zero, and heavy snowfall. Paris, on the other hand, has relatively mild winters with little in the way of snowfall, on average. Yet surprisingly, Paris is located farther north than Montreal. The reason for the difference in climate between these two cities is the moderating effect of the Gulf Stream.

The warming effects of ocean currents are not confined to Europe. In general, northward moving warm ocean currents, such as the Gulf Stream off the east coast of North America and the Japan current off the east coast of Asia, keep the polar regions of the Northern Hemisphere much warmer than they would otherwise be.

Scientists consider this an important key to the comings and goings of the ice ages. Milankovitch cycles can set the stage for an ice age by reducing the amount of solar energy that reaches the polar regions. Continents provide the land over which glaciers can spread. But scientists think that the oceans may act as an important trigger that can send the global climate from a cool state into a downright cold one.

This can happen if the ocean currents shift, break down, or are interrupted somehow. There are three main forces that drive the currents of the ocean. The first is wind. Air blowing along the surface of the oceans pushes water along underneath. While winds shift in direction from hour to hour with the passing of weather systems, over long periods of time winds tend to blow from the same direction more often than not. This is particularly true in the tropics, where wind flow is almost always from the east. Tropical easterly winds push water along toward the west. Outside the tropics, winds mostly blow from the west. These westerly winds push water along toward the east.

The second force that acts to shape ocean currents is the position of land masses. An easterly flowing current, for example, must change direction if it encounters a land barrier such as a large island or continent. In general, eastward flowing currents tend to turn toward the north as they encounter land masses. As they drift farther and farther north, they begin to move out of the tropics and into the middle latitudes of the earth. Here, westerly winds deflect the northward flowing ocean currents, sending them more toward the northeast.

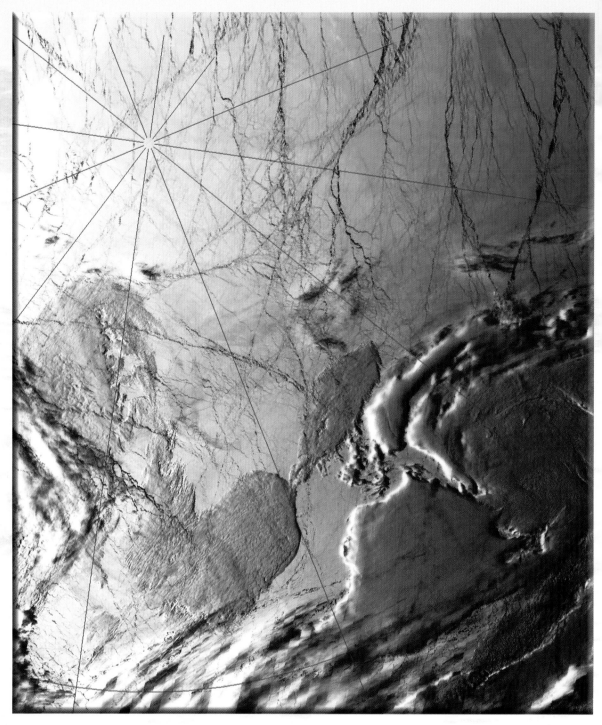

Milankovitch cycles can set the stage for an ice age by reducing the amount of solar energy that reaches the polar regions, such as the North Pole, pictured here.

The third force that causes ocean water to move is the density of water molecules. Ocean water density is controlled by water temperature and by the amount of salt in the water, also known as the water's salinity. The lower the amount of salt, the denser the water becomes. If the water becomes dense enough, as occurs in the North Atlantic near the end of the Gulf Stream's path, it can sink to the bottom of the ocean and travel back toward the equator as a cold, deep underwater current.

The presence of land masses changes the flow of ocean currents, which influence the world's climate.

Changing one or more of these driving forces can alter the way ocean water circulates around the planet. One way in which ocean currents can become disrupted is through a long-term change in the weather. Milankovitch cycles can reduce the amount of solar energy reaching the polar regions, causing these areas to become colder. The colder the polar regions become, the stronger the difference in temperature from north to south across the planet. And as we've seen, this north-south temperature difference helps drive the earth's weather machine. Change the way temperature varies across the planet and you change weather

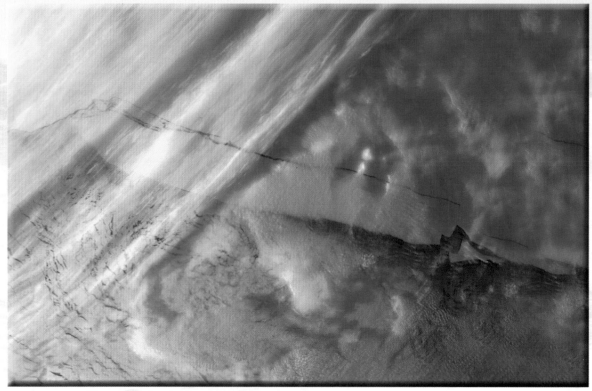

This is a picture of a massive iceberg, known as B-15, that broke off the Ross Ice Shelf in Antarctica in March 2000.

patterns. Changing weather patterns means shifting winds, which can alter ocean currents.

Another way that ocean currents can shift or break down is through the movement of land masses. The earth's crust, on which the oceans and continents sit, is divided into enormous, jagged fragments called plates. These plates are constantly in motion. Two hundred million years ago, for example, the Atlantic basin was a narrow fraction of its current expanse. South America hugged up against Africa, while western Europe butted up against eastern North America. Millions of years from now, the arrangement of land masses and oceans will be very different. And since the edges of land masses serve as deflecting boundaries for ocean currents, shifting land masses result in shifting ocean currents. Should ocean currents change in such a way as to lessen the amount of warm water flowing toward the poles, the polar regions would become colder still.

What's more, the arrangement of land masses can make it easier for an ice age to occur. Glaciers spread much more easily over land than over water. When land masses are positioned close to the poles, as they currently are in the Northern Hemisphere, they can serve as the spawning ground for ice sheets. In fact, the vast ice sheet covering Greenland today is a remnant of the enormous glacier that covered much of North America 20,000 years ago. On the other hand, where land masses are far removed from the poles, as they currently are in the Southern Hemisphere, ice sheets are less able to develop.

Another factor that influences climate is the amount of carbon dioxide in the air. Carbon dioxide is a greenhouse gas, meaning that it helps to regulate the temperature of the atmosphere by absorbing radiation from the earth. In general, the more carbon dioxide in the atmosphere, the warmer the planet; the less carbon dioxide in the atmosphere, the colder the planet. Carbon dioxide is cycled through the oceans, land, and plants through natural processes that occur over millions of years. If, over time, these processes cause carbon dioxide levels to drop, then global temperatures may fall and an ice age becomes more likely. However, scientists are unsure whether

decreasing amounts of carbon dioxide help to cause an ice age or are an aftereffect of an ice age.

The progression of events that leads to an ice age is extremely complex. While Milankovitch cycles, carbon dioxide levels, and an agreeable arrangement of land masses can stack the deck in favor of an ice age, ocean currents and weather patterns are the wild cards. Scientists describe the behavior of air and water currents as "chaotic," meaning they are basically unpredictable. Warm or cold water changes the temperature of the air above, and the blowing wind drives the water beneath. These two vast natural forces can coexist in a stable relationship for hundreds or thousands of years. Ocean currents flow along year after year, circulating water between the tropics and poles. Weather patterns vary from day to day and season to season, but remain within the general boundaries of the long-term climate.

However, the slowly changing Milankovitch cycles and creeping land masses gradually strain the system. Over centuries, the polar regions receive a shade less sunshine, growing imperceptibly colder as the orbit of the earth shifts little by little. Oceans widen or contract a couple of centimeters per year. At some point, when the temperature difference between the poles and the tropics reaches a certain threshold, weather patterns suddenly change. Or, when widening and contracting oceans reach a certain width, ocean currents suddenly shift. Together, the oceans and the atmosphere jump into a new relationship. Occasionally, over the billions of years of the earth's history, this ocean-atmosphere relationship has been a chilling one, leading to a long, global, winterlike season. Ocean currents stop circulating warmer air northward toward the poles. Energized weather systems carry more and more moisture over the continents, wringing themselves out as great snowstorms. Cooler summers melt less and less of the snow, until one year the snow doesn't melt during the summer and begins to harden into ice. Over centuries, with snow piling up in layers each year, the ice grows and begins to expand across the land. The bright white glacial ice, with its high albedo, reflects more and more sunlight back into space away from the earth. As the earth absorbs less and less sunlight, it grows colder, aiding the expansion of the ice sheets. The more they grow, the more sunlight gets reflected and the colder the atmosphere becomes. Thus, over thousands of years, an ice age descends on the planet.

4 Ice Ages of the Future

It is difficult to picture the effects of an ice age on our civilization tens of thousands, or even millions, of years in the future. Will humans have migrated to other planets by then? Will we have invented technology that controls the atmosphere—a global thermostat of some kind that we can fine-tune to prevent an ice age from even occurring? Will there still be towns and cities? After all, civilization as we know it has existed for only the last few thousand years. Science and technology are only a few hundred years old.

So as we imagine such a distant ice age, we must draw from what we know about past ice ages. We may imagine a time when Milankovitch cycles send a deep chill across North America and northern Asia. Slowly, almost without notice, temperatures drop year after year. Winters become

A future ice age could completely cover the Great Lakes under a sheet of ice.

colder. Outbreaks of Arctic air spread farther and farther southward with each passing decade, sending temperatures plummeting. Certain trees and plants, used to living in milder climates, begin to die. Over the passing centuries, fruit trees vanish from the Florida landscape as wintertime temperatures regularly plunge well below freezing. Snowstorms begin to frequent the Gulf Coast, whitening the beaches where people once flocked among the palm trees.

The Great Lakes have become locked in ice. The ice grips the lakes earlier each fall and melts later and later each spring. One summer finally arrives when the lake ice never completely melts, a prelude to the complete disappearance of the Great Lakes underneath a permanently frozen surface. Likewise, the Mississippi River becomes choked with ice farther and farther south each winter with each passing century. Eventually, ice begins to flow past the location of New Orleans. Across the central part of North America, where cities like Denver and Chicago stand today, winter conditions become more severe over time. Eventually, air temperatures routinely plunge below -50°F. Summer grows short and cold. Snow becomes more and more common in May, then in June.

Weather conditions will become harsher as a new ice age approaches.

Blizzards strike with increased vigor and frequency along the East Coast of the United States as colder wintertime air sharpens the temperature difference between the land and the still-warm Gulf Stream offshore. Energized winds in the atmosphere spin up great coastal storms that dump snow by the foot from the Appalachians northward. And on the West Coast of North America, immense Pacific storms crash inland, flooding the valleys and lowlands and layering the mountains with snow deep enough to cover trees. As ice age conditions spread across the planet, the dry, western part of North America will become wet. Deserts will bloom. Lakes will expand and overflow. Rivers will rush wild through the summer as frequent rains and massive amounts of melting snow send water in torrents toward the sea. In the mountains, glaciers will grow and spread down into the lowlands, the plowing ice slowly knocking down forests.

As an ice age grips the planet, thousands of years in the distant future, we look to the polar regions for the coming of the ice sheets. The great mass of ice covering Greenland today will overflow the island, spilling into the nearby ocean water. As the ice expands, it will connect with the growing mound of ice and snow over northern North America. One summer, far in the distant future, the winter snowpack will not completely melt as it has every year throughout history. As the years and decades advance, the snow will deepen, hardening into ice. The ice will thicken and expand, creeping southward through the great forests of North America. Trees will die by the thousands, becoming engulfed in a rising tide of frozen white.

A similar scenario will unfold across Asia. Glaciers expanding out of Scandinavia will spread across modern-day Russia. The Baltic Sea and the North Sea will fill with icebergs, then become completely covered in ice. The British Isles will turn into a glacial landscape. Other great glaciers will spread out of the

While the rest of the world will be drastically altered by an ice age, the tropics may not change very much.

Himalayas. In the Southern Hemisphere, ice will expand more slowly, owing to the relatively small amount of land close to the South Pole. But the Antarctic ice sheets will grow, sending icebergs floating farther and farther north toward the equator.

Meanwhile, far from the crumbling edge of the great ice sheets, the tropics will probably continue to bask in relative warmth. In fact, scientists now think that average tropical ocean water temperatures during the last ice age may have been higher than today as ocean circulation collapsed, shutting off cooling currents from the poles.

This drastic difference in temperature over such a short distance, from the balmy tropics to the frigid ice sheets, will set the earth's weather machine into high gear. Winds will accelerate and storms will strengthen in much the same way that they do today as summer transitions into autumn, then winter. In a future ice age, the

These penguins might be forced to migrate if a new ice age occurs.

atmosphere will be locked in an energized, winterlike pattern all year round.

The energized atmosphere will also contribute to the expansion of the ice sheets. During a future ice age, energized wind currents will carry large amounts of moisture northward from the still-warm tropical oceans, dumping it as snow over the ice sheets. This will help the ice sheets to grow, intensifying the temperature contrast between pole and equator. The stronger the temperature contrast, the stronger the winds will become and the more moisture will be carried northward to fall as snow.

Animals and plants will either adapt to the colder weather or they will become extinct. Where the ice doesn't scour the land of vegetation, whole ecosystems—collections of plants and animals living in connected relationships—will migrate southward. Where today stand thriving fields of corn, wheat, and soybeans across the heartland of the central United States, in a distant ice age there will stand forests of spruce and perhaps herds of caribou. Fish will migrate southward toward warmer water in advance of the spreading ice. Perhaps one day, in a distant ice age, drifting ice-bergs will carry penguins off the coast of eastern Florida.

The ice age of the distant future, like ice ages of the distant past, will reshape the land. Hills and smaller mountains will be flattened by hundred- or thousand-foot deep layers of ice. Huge chunks of ice will gouge holes in the ground while house-sized boulders will be lifted and carried along by the advancing glaciers. Perhaps the Great Lakes and other large lakes across northern Europe and northern Asia will disappear underneath the ice, vanishing forever as the glaciers fill them with rocks and sediment. The Mississippi, Ohio, and Missouri Rivers will shift and expand into a vast network of braided channels, carrying meltwater, sand, and rocks from the edge of the glaciers southward across the continent. High winds blowing across these outwash plains will lift great clouds of dust into the sky.

These are all possible images from an ice age of the future. A future ice age will take tens of thousands, or millions, of years to evolve. It may last for tens of thousands of years, then give way to slowly warming conditions over hundreds or thousands of more years. This has been the record of ice ages in the past, discovered by scientists looking at cores from ice, oceans, and lakes.

But scientists have also discovered startling evidence of climate change that occurs much more rapidly. In the next chapter, we will examine these rapid climate fluctuations and their possible relationship to the warming trend that scientists think is currently occurring.

5 Cold Snaps

It seems strange to think about ice ages during a time when the earth is warming. Scientists estimate that the average temperature of the earth increased by 1°F over the twentieth century. While one degree seems small, it's significant considering that the earth has warmed by only five to nine degrees since the depths of the last ice age 20,000 years ago. The warmest decade of the twentieth century, and perhaps for the last 1,000 years, was the 1990s. And scientists think that the rate of warming is accelerating.

The cause of the warming has been traced to increasing amounts of greenhouse gases in the atmosphere. Greenhouse gases, including carbon dioxide, methane, and chlorofluorocarbons, are emitted from the burning of fossil fuels. Fossil fuels include

Burning fossil fuels, including the oil being extracted by this rig, causes global warming when the greenhouse gases these fuels produce absorb the Sun's radiation.

natural gas, coal, and oil—the primary sources for electricity. As more and more fossil fuels are burned in power plants and vehicles, more greenhouse gases are released into the atmosphere. There they efficiently absorb radiation emitted by the ground below, radiation that would otherwise escape into space. This radiation is converted into molecular energy. As the atmosphere gains energy, it emits more radiation to the ground below, causing the ground to warm. The more greenhouse gases, the more energy is absorbed by the atmosphere and the warmer the earth becomes.

Some scientists think that global warming—the same global warming that is melting glaciers around the world and ice caps in Antarctica and Greenland—could lead to a rapid, drastic change in global climate. In one scenario, this climate change will take the form of a global cold snap. While it would not result in a full-blown ice age, it could potentially be very disruptive and damaging as temperatures plunge in certain parts of the world. How could this happen? It has to do with the way global warming affects the oceans.

Scientists are particularly concerned about the effects of global warming on that part of the ocean circulation that flows through the North Atlantic. Here, dense, salty water sinks to the bottom of the sea and begins a slow southward drift back toward the equator. Eventually, this deep ocean current meanders into the South Atlantic, across the Indian Ocean, and into the Pacific, where it rises to the surface. The surface current then retraces the route, back through the Indian Ocean and South Atlantic, across the equator, and into the Gulf of Mexico, where it becomes the Gulf Stream. The entire trip takes over 1,000 years. The sinking water in the North Atlantic is a critical part of this circulation.

Global warming may force changes in the North Atlantic in two ways. The first is by changing weather patterns. Scientists agree that one of the effects of global warming is an increase in the moisture content of the atmosphere. This moisture comes mostly from water evaporated from the oceans. In general, the higher the temperature, the higher the rate of evaporation, and the higher the atmospheric moisture level. More atmospheric moisture means heavier rain, at least in certain regions of the world.

One of these regions may be the North Atlantic. More rainfall in the North Atlantic will cause the salt content of that part of the ocean to decrease, or become diluted. The less salt in the ocean, the less dense the ocean water. And ocean water density is what drives the sinking North Atlantic current.

There's another way that the salt content of North Atlantic water may drop. As global warming continues, higher temperatures melt more icebergs and ice caps. For example, NASA scientists reported in the summer of 2000 that the edges of the Greenland ice cap were melting at a rate of three feet per year. The melting of ice adds more freshwater to the ocean. This contributes to the decline in salt content and water density, further increasing the chances that the critical sinking current in the North Atlantic may be disturbed.

Global warming may cause more evaporation of ocean water, which will lead to greater levels of moisture in the air, in turn causing heavier rainfall in certain parts of the world.

Interrupting this vital link in ocean circulation may have dire consequences. The Gulf Stream, which serves as the source of heat for the North Atlantic and Europe, may weaken or shift southward. Without the warming effects of Gulf Stream waters, some scientists think that average temperatures across Europe would drop by 9 to 18°F. This kind of change may be catastrophic. One of the first victims would be agriculture. Farms across Europe, including the great vineyards of France, Germany, and Italy, would be devastated as the growing season shrinks to a fraction of its former length. More frequent snowstorms would bury cities each winter, snarling transportation. The climate of the British Isles might become more like Norway or even Iceland.

The repercussions from an interruption of ocean circulation would not be confined to Europe. Weather patterns are connected to one another, so that a major change in weather in one region of the world can cause problems on the other side of the world. If the Gulf Stream shifts south or weakens from an influx of freshwater into the North Atlantic, weather patterns would change in many parts of the Northern Hemisphere. With the Gulf Stream unable to transport heat toward the pole, the temperature contrast between the North Pole and the equator would intensify. And as we've seen, this may result in stronger storm systems and perhaps more severe winters— especially across northern parts of North America and Asia.

Unlike an ice age, this kind of global cold snap may occur suddenly, in a decade or less. Ice ages, on the other hand, take tens of thousands, or millions, of years to evolve. Evidence for cold snaps comes from the ice cores drilled by scientists in Greenland. One of the more notorious of these long-ago cold snaps has been named the Younger Dryas. The name comes from a species of Alpine flower that reappeared in England as the climate there suddenly shifted. The Younger Dryas occurred as the earth was slowly warming out of its last ice age. Suddenly, around 12,000 years ago, the climate swung back into a

deep chill, lasting for nearly 2,000 years. Glaciers expanded once again as snowstorms and frigid air spread toward the equator.

Scientists think that the Younger Dryas was triggered by a catastrophic flood of freshwater into the North Atlantic. For thousands of years, ice sheets had blocked meltwater from flowing into the St. Lawrence seaway, a natural drainage system extending from the Great Lakes into the North Atlantic. Meltwater dammed up behind the ice sheets was let loose in a cataclysmic torrent as the glaciers finally retreated around 12,000 years ago. This enormous discharge of low-density freshwater swamped the North Atlantic and shut down the ocean circulation, plunging the earth back into ice age conditions in a matter of decades. It took over 1,000 years for the ocean currents to reestablish themselves and begin the global warming anew. But when the cold snap ended, it did so just as fast as it had begun. Ice core data show that global temperatures at the end of the Younger Dryas increased by as much as 10°F and precipitation decreased by as much as 50 percent over a period of fifty years.

Cold snaps, therefore, pose the greatest threat for colder, ice age–like conditions in the coming centuries. The cold snap scenario is nearly impossible to foresee, however. This is because it depends on the atmosphere and the oceans—two chaotic and essentially unpredictable natural systems. Scientists are unsure exactly how much climate change is needed to trigger the shutdown of ocean circulation. No scientist can even say for sure when or even if such a drastic climate change will occur. But the possibility exists and is supported by data from ice cores and records.

Conclusion

The earth has experienced rapid swings in climate throughout its history, in addition to the slow comings and goings of the ice ages over tens of thousands of years. Human civilization has evolved during an especially tranquil period in the earth's climate, but now civilization has reached a point where it has begun to influence climate. Global warming, caused by increasing amounts of greenhouse gases in the atmosphere, poses a variety of threats to the planet. Surprisingly, one of these threats is to the ocean circulation that plays a role in the comings and goings of ice ages. The lesson from ice cores is that the earth's climate has not always been as tranquil as it has been during human history. If global warming continues as many scientists expect, it may reintroduce us to the more unstable, fickle side of climate, and even possibly someday send a chill across the earth, giving us a taste of what the weather was like during our remote past. The future may not be as warm as we think.

Glossary

albedo The amount of solar radiation, or sunlight, reflected by a surface, compared to the amount that it absorbs.

climate The average weather conditions over a long period of time, generally decades or more.

evaporation The process whereby liquid water changes into invisible, gaseous water vapor.

fossil fuel Any fuel made from the decayed remains of ancient plant life. Includes coal, natural gas, and oil.

glacier   A very slowly moving river or sheet of ice.

global warming   The warming of the planet due to increasing amounts of greenhouse gases in the atmosphere.

greenhouse gas   Any gas that efficiently absorbs outgoing radiation from the earth. The main greenhouse gases are water vapor, carbon dioxide, methane, nitrous oxide, chlorofluorocarbons, and ozone.

Gulf Stream   A great ocean current, beginning in the Gulf of Mexico south of the United States, then curving around the southern tip of Florida and northeastward into the Atlantic.

ice age   A period of time in the earth's history when the global climate was especially cold, causing vast ice sheets and glaciers to expand across continents.

ice core   A long tube of ice drilled out of an ice sheet or glacier. Studying ice cores enables scientists to learn about changes in the earth's climate over long periods of time.

isotope A variety of atom of a particular chemical element. Isotopes of an element vary in their weight.

Milankovitch theory A theory named after Milutin Milankovitch, a Serbian mathematician, that accounts for changes in global climate, including ice ages. Slow variations in the earth's orbit around the Sun over tens of thousands of years are thought to affect the way solar radiation is distributed around the planet.

paleoclimatology The study of climates long ago in the earth's history.

radiation Energy in the form of invisible electromagnetic waves that travel at the speed of light.

salinity The amount of salt in a substance or liquid.

Younger Dryas The name given to a sudden global cold snap occurring from 12,000 to 10,000 years ago.

For More Information

American Meteorological Society (AMS)
45 Beacon Street
Boston, MA 02108-3693
(617) 227-2425
Web site: http://www.ametsoc.org/AMS
The AMS is a professional meteorological organization in the United States.

Climate Prediction Center (CPC)
World Weather Building
5200 Auth Road, Room 800
Camp Springs, MD 20746
(301) 763-8000
Web site: http://www.nnic.noaa.gov/cpc
The Climate Prediction Center's Web site offers information on current global temperature trends.

Greenland Ice Sheet Project 2
GISP2 Science Management Office
Climate Change Research Center
Institute for the Study of Earth, Oceans, and Space
University of New Hampshire
Durham, NH 03824
(603) 862-1991
Web site: http://gust.sr.unh.edu/GISP2

Intergovernmental Panel on Climate Change (IPCC)
c/o World Meteorological Organization
7 bis Avenue de la Paix, C.P. 2300
CH-1211 Geneva 2
Switzerland
+41-22-730-8208
Web site: http://www.ipcc.ch
The IPCC issues scientific reports on global warming and
its effects.

Weatherwise Magazine
Heldref Publications
1319 18th Street NW
Washington, DC 20036-1802
Web site: http://www.weatherwise.org
This is a magazine about all things weather.

For Further Reading

Bailey, Ronald H., and editors of Time-Life Books. *Glacier.* Alexandria, VA: Time-Life Books, 1982.

MacDougall, J. D. *A Short History of Planet Earth: Mountains, Mammals, Fire, and Ice.* New York: Wiley, 1986.

Stevens, William K. *The Change in the Weather: People, Weather, and the Science of Climate.* New York: Delacorte Press, 1999.

Index

About the Author

Paul Stein has a B.S. in meteorology from Pennsylvania State University. He has eight years' experience as a weather forecaster, most recently as a senior meteorologist for the Weather Channel. Currently he develops computer systems and software that display and process weather-related data.

Photo Credits

Cover © Pictor: Stephens Passage, Alaska.
Cover inset © NASA JPL: space radar image of Oetzal, Austria, where the Alps converge at Switzerland, Italy, and Austria.
Front matter and back matter © Pictor: Disco Bay, Greenland.
Introduction background © DigitalVision: snow-covered mountain peaks.
Chapter 1 background © Pictor: Antarctica.
Chapter 2 background © Weatherstock: trees covered with snow.
Chapter 3 background © Artville Weatherstock: ocean waves and currents.
Chapter 4 background © Corbis/Visions of Nature: ice.
Chapter 5 background © Weatherstock: satellite image of hurricane.
P. 10 © Pictor; p. 11 © Jacques Descloitres, MODIS Land Science Team; p. 12 © Superstock; p. 15 © UCLA/AP/Worldwide; pp. 18, 39, 49 © Artville Weatherstock; p. 21 CERES Instrument Team, NASA Langley Research Center; p. 23 © Reto Stockli, Nazmi El Saleous, and Marit Jentoft-Nilsen, NASA/Goddard Space Flight Center (GSFC); p. 24 © Robert Simmon,

NASA/GSFC; p. 30 © Weatherstock; pp. 31, 32, 38, 50–51 © SeaWiFS Project, NASA/GSFC and Orbimage; pp. 41, 42, 46 © Pictor.

Series Design and Layout

Geri Giordano